THE AGIUS
Evolution Garden

Richard Wilford
Sharon Willoughby

Kew Publishing
Royal Botanic Gardens, Kew

Introduction

The Agius Evolution Garden occupies a corner of the Royal Botanic Gardens, Kew that has been a systematic garden (an area illustrating plant classification) for over 170 years. As our scientific knowledge has increased and classification systems have changed, so has the arrangement of plants in this area. In 2019 this part of Kew was completely redesigned to create a garden based on the latest research into plant relationships. This classification reveals a new understanding of how plants are related to each other and how they have evolved over more than 400 million years, from the first plants to colonise land to the huge diversity of beautiful flowering plants that exist today.

Traditionally, plant species were grouped into genera and families according to their physical characteristics (morphology). Now, the study of plant DNA (deoxyribonucleic acid) has allowed scientists to uncover how closely different plants are related and how they have evolved from their recent ancestors, resulting in a more accurate and robust classification system. DNA evidence has shown us some surprises. Some plants that we once thought were unrelated because they didn't share physical traits, such as plane trees (*Platanus*) and the lotus flower (*Nelumbo*), are, in fact, very closely related.

The layout of the Evolution Garden reflects Kew's work in the global effort to uncover the Tree of Life, enabled by recent advances in DNA sampling technology. The classification that inspired the design of this garden was created by the Angiosperm Phylogeny Group (APG), which includes several researchers from Kew. In 1998, the APG created the first DNA-based classification of plants. Several updates have been published subsequently, leading up to APG IV in 2016.

The Evolution Garden is divided into eight sections, separated by low yew hedges to form garden 'rooms'. Closely related plants are grouped into families, and related families are grown together in a garden room. Each room represents a different branch of the Tree of Life and is home to a range of plant families that have evolved from a common ancestor.

[opposite]
The Evolution Garden.

History of the Site

Before 1730, the area of Kew now partly covered by the Evolution Garden belonged to the Capel Family. The northern part of the land was an orchard, and the southern end was divided into four long borders, running north to south.

From 1730 to 1846, it was part of the 14-acre Royal Kitchen Garden. Originally, this garden supplied fruit and vegetables to the household of Prince Frederick and Princess Augusta. It was part of a network of gardens, including Hampton Court, Kensington Palace, and Windsor Castle that fed the royal households around London until the reign of Queen Victoria.

Between 1846 and 1867 it was known as the Herbaceous Ground. Sir William Jackson Hooker, Kew's first director, arranged it into a series of round parterres and interlocking beds. Its planting reflected the classification published in *Genera Plantarum* (1789) by the French botanist, Antoine Laurent de Jussieu.

The series of rectangular beds long known as the Order Beds were first laid out between 1867 and 1870 under the direction of Sir Joseph Dalton Hooker, Kew's second director. Between 1862 and 1883, Hooker worked with another Kew scientist, George Bentham, to develop a plant classification known as the Bentham and Hooker System. Hooker designed the garden to reflect this classification. Each rectangular bed was dedicated to one family, with larger families occupying more than one bed. As in the Herbaceous Ground, the plants grown were mostly hardy herbaceous perennials and annuals. The Rose Pergola along the main paths was built in 1959 to celebrate the 200th anniversary of Kew Gardens.

The Order Beds were replanted in 2010 to show some of the early changes that DNA technology brought to our understanding of plant relationships. It reflected the classification of plants in APG version III, published in 2009. It was renamed the Plant Family Beds, but the arrangement of rectangular beds and narrow grass paths stayed. As we discovered more about plant relationships, it became apparent that this design no longer represented current plant classification, so plans for a new layout were made.

[opposite top]
1837 plan of the Royal Kitchen Garden, outlined in red. The Evolution Garden now occupies the southern end.

[opposite bottom]
The Order Beds, showing the arrangement of rectangular beds that were first laid out between 1867 and 1870.

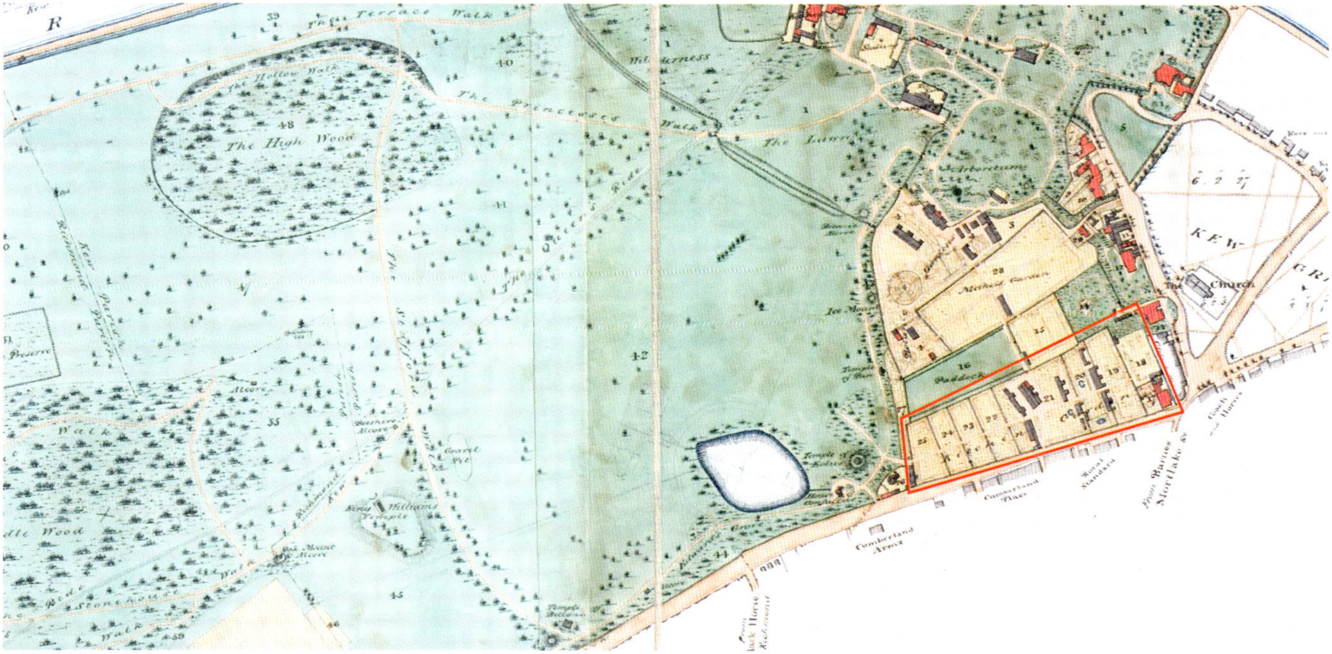
Terrace Walk
The High Wood
Wilderness
The Princesses Walk
Arboretum
Melon Garden
Paddock
Church
KEW
Cumberland Place
Cumberland Arms
Royal Standard
Mortlake St
Gravel Pit
Stonehouse

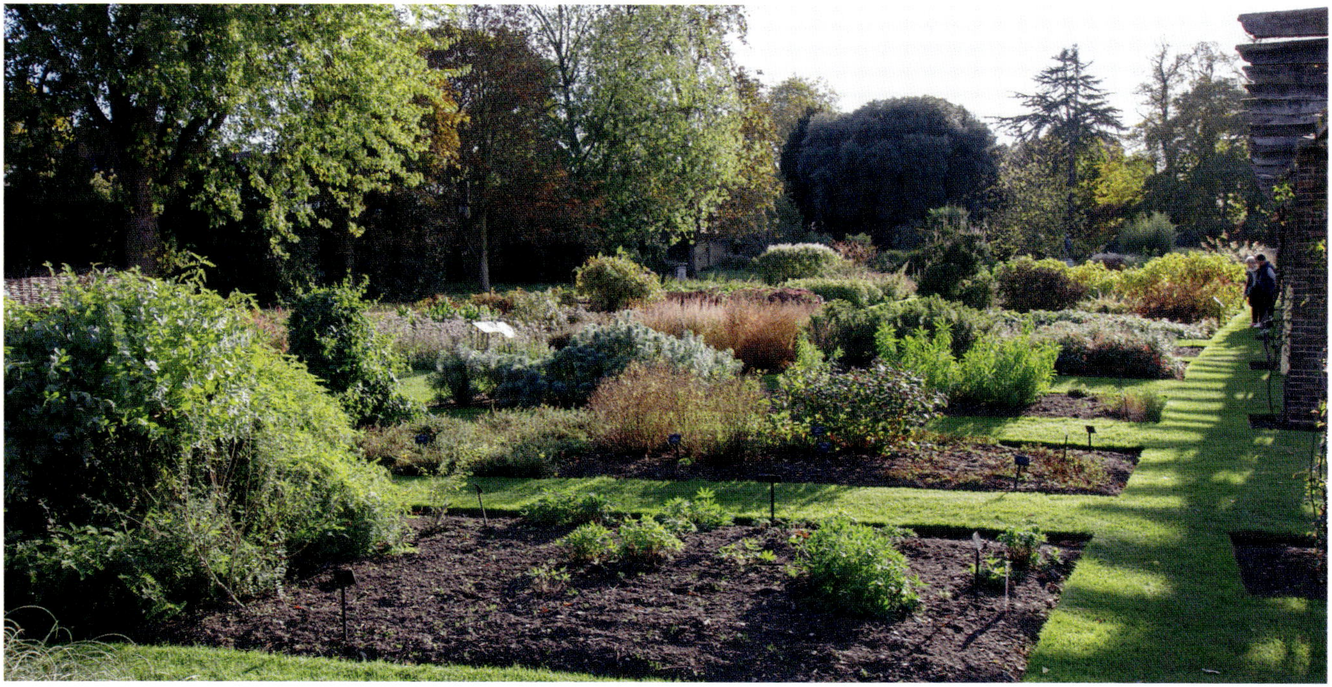

The evolution of plants

The further we look back in time, the harder it is to find evidence from the fossil record. When new fossil evidence is found it can alter our understanding of how life evolved and the timescale over which it happened.

Geologists estimate that planet Earth formed some 4.6 billion years ago. Scientists propose that the first sparks of life occurred in Earth's primordial soup, around 3.8 billion years ago. The earliest evidence of plant-like organisms capable of photosynthesis emerges in the fossil record 2.5 billion years ago, at the beginning of the Archaean eon. The simple algae that floated through the Earth's young oceans were some of the very first forms of life.

Around 1.5 billion years ago, plants, animals, fungi and bacteria began to diverge from each other, creating the first branches of the Tree of Life. Plants went on to colonise the land between 480–500 million years ago and evolved the ability to produce seeds approximately 380 million years ago. The first flowers appear in the fossil record around 140 million years ago but it is likely that flowers evolved earlier than that, somewhere between 150 and 245 million years ago. To put this all into context, the first modern humans roamed the Earth a mere 200,000 years ago.

[above]
Plant fossils can be used, along with evidence from the molecular clock, to estimate the age of different plant groups. This fossil of the extinct *Leefructus mirus* from China, is at least 122.6 million years old and shares striking similarities with plants in the buttercup family (Ranunculaceae), indicating when this plant group first evolved. Image: David Dilcher/Ge Sun.

How DNA helps us understand evolution

Evolution is generally a gradual process that operates over an enormous time scale (typically millions of years). To create an accurate evolutionary history of life, we must understand when different species diverged from their shared ancestors. Fossil evidence can signpost when these events occurred, and scientists can now supplement this by looking to the genome and what is termed the 'molecular clock'.

The molecular clock technique is based on the theory that biological molecules, such as DNA, have a relatively steady rate of change over time, although this rate may vary greatly between groups of organisms. Some mutations in the DNA sequence are harmful and therefore tend not to persist across generations. But neutral mutations (those that do not affect the survival of an organism) can accumulate over time and become fixed throughout a population. When comparing DNA sequences between two species, the genetic similarity is, to an extent, proportional to the time since they last shared a common ancestor: few changes indicate it was recent, many changes much more distant.

Used by itself, the molecular clock only provides the degree of change, not concrete dates of divergence. For this, the clock needs to be calibrated using independent data, such as from fossils of known ages, to calculate the rate at which these molecular changes occur. The fossil record provides evidence of known evolutionary divergence events, such as the split between mammals and birds or flowering plants and conifers. By putting the age of these fossils together with the evidence from the molecular clock, the age of different groups can be estimated.

WHAT IS DNA?

Deoxyribonucleic acid (DNA) is the master code for life. It is a complex molecule that contains all the instructions for a living organism to survive, grow and reproduce. DNA exists in almost every living cell, from simple bacteria to the highly specialised cells in our bodies. It provides a blueprint of genetic information, unique to each organism.

DNA is made up of two strands of smaller units, called nucleotides, bound together in a double helix structure (a typical human cell contains c.6 billion nucleotides). The order of these nucleotides along the strands represents the DNA sequence of an organism with all the information it needs to function. Parts of the DNA sequence are organised into genes.

Genes code for different characteristics by telling cells how to make the proteins vital to the organism's function. The DNA is packaged into larger units called chromosomes, and all the chromosomes together make up the genome.

Plant DNA extraction in the Jodrell Laboratory at Kew.

The science behind the garden

Much like a family tree, the Tree of Life is made up of branches, each representing a line of descent. The Tree of Life depicts evolutionary history (phylogeny), illustrating relationships between all living things. In the past, the form of a plant (morphology) and other easily assessed characteristics such as the number of petals, leaf shape and chemistry, were used to assess relationships between species of plants and fungi. Today, sophisticated technology gives us the power to investigate genetics in detail, yielding DNA data that give a far more detailed picture of how and when plants evolved.

The Plant and Fungal Trees of Life (PAFTOL) project is one of Kew's most ambitious scientific endeavours. It aims to uncover the relationships between plant and fungal species based on the information held in their DNA. The Tree of Life is as crucial to biology as the Periodic Table is to chemistry, providing a roadmap for understanding life on Earth. The PAFTOL project is a vital step towards uncovering the Tree of Life for all plant and fungal species.

Kew's PAFTOL researchers aim to generate genomic (DNA) data for at least one species from each of the 14,000 flowering plant and 8,200 fungal genera. Many plant and fungal species needed for this research are already present in Kew's collections as living plants, seeds, or previously extracted DNA samples. The PAFTOL project also recovers DNA from Kew's seven million herbarium specimens, some of which were collected over 200 years ago.

The more we understand the living world, the better equipped we are to meet current challenges including biodiversity loss, climate change and the spread of pests and diseases. By revealing the genetic links between plant and fungal species, a completed Tree of Life will enable scientists to investigate species closely related to those known to have useful properties. Ultimately, this greater understanding will help us to better conserve the amazing diversity of plants and fungi on Earth.

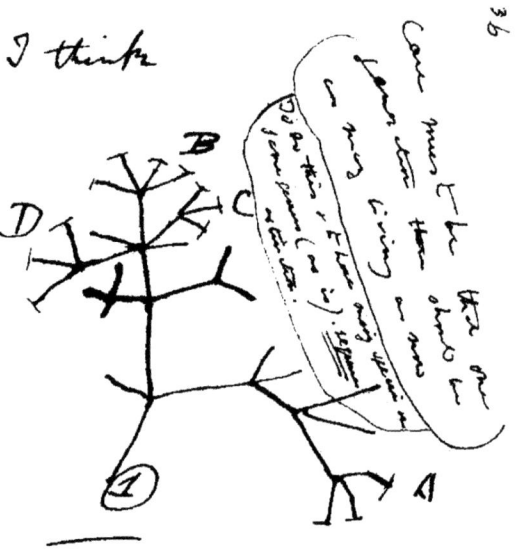

[above]
Charles Darwin's 1837 sketch of an evolutionary tree from his first notebook on 'Transmutation of Species'.

THE ANGIOSPERM PHYLOGENY GROUP

Historically, plant classification has been led by one or a few individuals, with many debates about which features should be used to classify plants. The Angiosperm Phylogeny Group (APG) has transformed this approach by drawing on the work of scientists worldwide to tell a cohesive story about the Tree of Life. The APG system is based on an explicitly analysed data set, rather than an intuitive interpretation of available data.

In 1998, Kew's senior research professor, Mark Chase, together with 41 co-authors, published a groundbreaking scientific paper analysing the DNA of 500 flowering plants. The resulting family tree revealed many surprising relationships. Scientists across the world have built on this understanding, sequencing the genomes of hundreds of plants.

The most recent revision, APG IV, was published by ten compilers from six countries in 2016, as the result of an international workshop hosted by Kew. With new plant genomes being sequenced every day, APG IV will not be the final version – but the task of reorganising plant collections is becoming increasingly achievable now they reflect evolutionary relationships between plants.

[right]
One early finding from analysing plant DNA was that roses (*Rosa*) and the common stinging nettle (*Urtica dioica*) are more closely related than previously thought.

Designing the Evolution Garden

The design of the Evolution Garden has three main aims. Firstly, it shows the latest classification of flowering plants, based on DNA analysis, and highlights the role of Kew's scientists in this research. There was also a desire to make the garden more aesthetically pleasing, to showcase the beauty of flowering plants, while still maintaining the scientific integrity of a systematic garden. Thirdly, it is vital that the layout makes sense to Kew's visitors and the arrangement of plants, along with interpretation, tells the story of plant evolution.

It was clear at the beginning of the design process that the arrangement of rectangular beds had to go. The focus had to move away from the 'one family per bed' concept to a more fluid layout that represented the related groups of plants and families that had been revealed in the new classification. The design had to represent the Tree of Life in living form. Each garden room in the Evolution Garden represents a branch or cluster of branches, from the Tree of Life. These groups are separated by yew hedges, creating the garden rooms. As the hedges grow, they will be clipped to create a neat and clear division so as you pass through a yew hedge, it is obvious that you are moving from one group to another.

The site of the Evolution Garden is bounded by a long brick wall on one side, the boundary of Kew Gardens along the other and is bisected by the rose pergola. The new layout reflects this angular geometry, with paths leading off the rose pergola into each room and cutting through the yew hedges. Solid oak benches are placed in each garden room, on brick pavers. All steps have been removed and

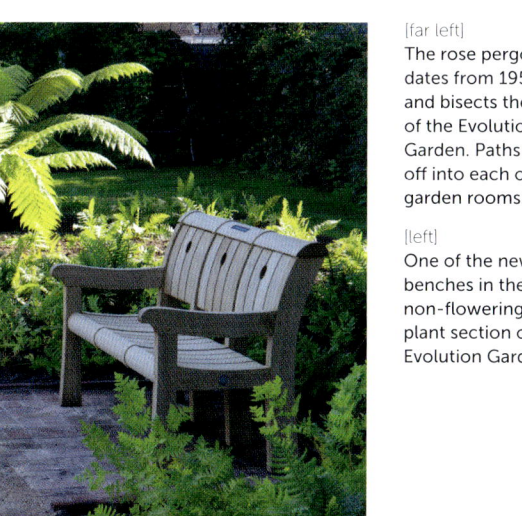

[far left]
The rose pergola dates from 1959 and bisects the site of the Evolution Garden. Paths lead off into each of the garden rooms.

[left]
One of the new benches in the non-flowering plant section of the Evolution Garden.

the paths are made with firm but porous self-binding gravel, so the site is fully accessible.

The planting beds in the Evolution Garden are much larger than the previous rectangular beds. This allows greater flexibility in the planting design. It means that less decorative plant families need only be represented by a few species, while families that contain a large number of attractive plants can cover a wider area and sometimes drift across the paths, giving flow to the design and creating a more immersive experience.

Species in one family are still clustered together. For example, all the plants in the iris family (Iridaceae) are grown together in one part of the monocot garden room. Closely related families are also grouped together, one example is the planting bed devoted to the peony family (Paeoniaceae), saxifrage family (Saxifragaceae) and stonecrop family (Crassulaceae). All of these families are grouped in the order Saxifragales. This demonstrates one of the early findings from DNA analysis that peonies are not closely related to buttercups, which have a similar looking flower structure, but are more closely related to saxifrages.

The interpretation in the Evolution Garden is as important as the plants. Each garden room has a cluster of panels that explains what is grown in that room. There are also panels about DNA and APG, as well as an overview of plant evolution. At each of the three entry points to the garden are panels showing the Tree of Life diagram and the layout of the garden. In addition, various science stories are highlighted throughout the garden, focussing on particular plants or families. The aim is to bring the science of plant evolution to life by highlighting the involvement of Kew's scientists in this fascinating, and ongoing, voyage of discovery.

[right]
The flower of *Paeonia daurica* subsp. *mlokosewitschii.*

[far right]
The Tree of Life panel at the southern entrance to the Evolution Garden.

DESIGNING THE EVOLUTION GARDEN   11

Planting the Evolution Garden

For over 170 years, the systematic garden at Kew, whether called the Herbaceous Ground, the Order Beds or the Plant Family Beds, has shown plant classification using mostly hardy herbaceous perennials and annuals. This tradition has been followed in the Evolution Garden. Not all families are represented, as many contain huge trees, aquatic plants or tropical species that cannot grow outside in the UK, but a wide range of over 700 different plant taxa (species, subspecies or cultivars) are planted here. Some woody plants have been introduced to increase the diversity of genera and families, such as the flowering cherry (*Prunus* x *subhirtella* 'Autumnalis Rosea') in the rose family (Rosaceae), an olive tree (*Olea europaea*) in the olive and jasmine family (Oleaceae) and heavenly bamboo (*Nandina domestica*) in the barberry family (Berberidaceae).

About a third of the plants in the Evolution Garden, including the olive tree, were saved from the Plant Family Beds before they were emptied. Many were lifted, kept in trays of soil over winter and replanted in the new design, while others were propagated by seeds or cuttings taken the preceding summer. By the end of 2018 the beds were clear, ready for landscaping. From January to May 2019, the irrigation system was installed, the new paths marked out and laid, and the path running beneath the Rose Pergola resurfaced. As the landscapers finished a section, Kew's horticulture team moved in to start planting. First to go in, during March, were the yew hedges. Then from April to July, the eight garden rooms were planted.

The plants are a mix of well-known garden plants and less common species. There are some cultivars, such as the white cone flower (*Echinacea purpurea* 'White Swan') and the oriental poppy (*Papaver orientale* 'Beauty of Livermere'), which are instantly recognisable to many gardeners. These more familiar plants help visitors identify a plant family and highlight some less obvious relationships by encouraging them to look more closely at nearby plants. It may not be immediately obvious, for example, that the large-flowered, brightly coloured oriental poppies are in the same family as the plume poppy (*Macleaya cordata*), which has tiny, straw-coloured flowers held on tall stems. An attractive garden will attract visitors, who will hopefully be intrigued by what they see and learn more about how plants have evolved and Kew's role in uncovering the Tree of Life.

[top left]
The Plant Family Beds were clear of plants by the end of 2018.

[below left]
Echinacea purpurea 'White Swan'.

[top right]
Planting the Evolution Garden began in April 2019.

[below right]
The plume poppy, *Macleaya cordata*, in the poppy family (Papaveraceae).

PLANTING THE EVOLUTION GARDEN   13

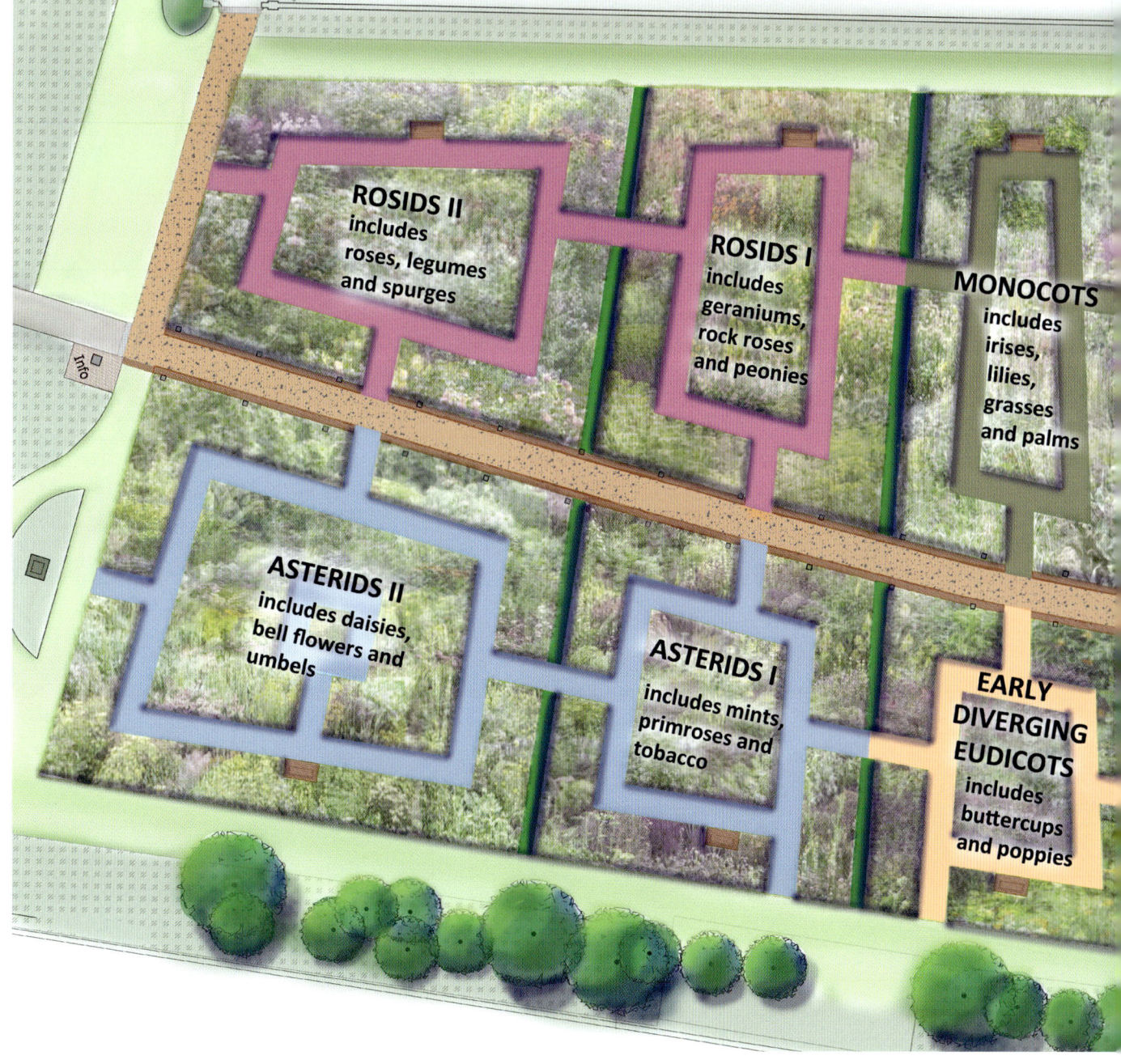

The Evolution Garden

MAGNOLIIDS includes magnolias, allspice and bay trees

NON FLOWERING PLANTS includes ferns and conifers

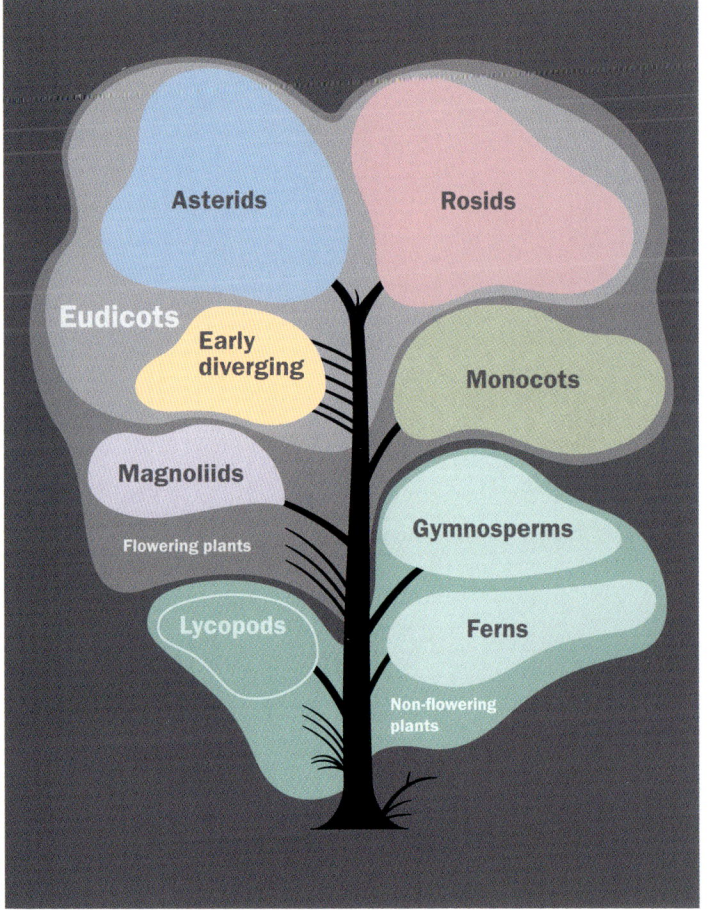

Plan of the Evolution Garden (left) with the path colours corresponding to the plant groups indicated in the Tree of Life (above).

Non-flowering plants

According to the fossil record, a remarkable thing happened around 484 million years ago, during the Upper Ordovician period. Plants, previously restricted to an aquatic life, began their conquest of land. In this garden room, the earliest examples of plants that evolved after the conquest of land are ferns like horsetails (*Equisetum*). The distant ancestors of these plants made a great evolutionary leap forward that allowed them to leave the water and exist in drier environments on land. These plants developed a vascular system of specialised cells allowing them to move water through their roots and stems. This innovation happened roughly 380 million years ago – the point

at which the ferns diverged from the main trunk of the Tree of Life.

The fossil record of the Carboniferous period (359 to 299 million years ago) tells us that forests were dominated by tree ferns and giant 40-metre-tall horsetails. Ferns are still abundant today. The giant horsetails, on the other hand, died out and gave way to woody plants with improved vascular systems and a new way of reproducing – seeds. Seeds were another great evolutionary leap, protecting plant embryos from dry environments on land. Seed bearing plants without flowers, collectively known as gymnosperms, are represented in the Evolution Garden by smaller conifers, the gingko and cycads.

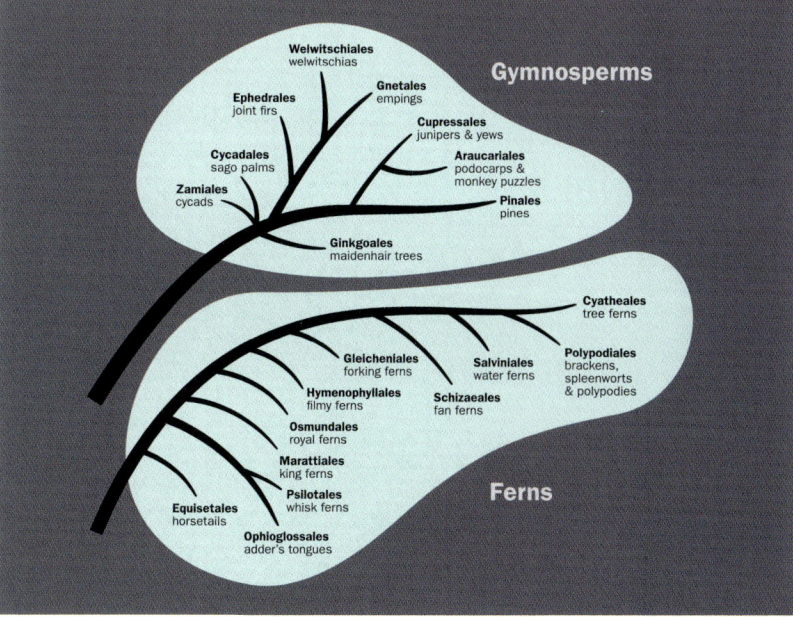

[opposite left]
The fern and gymnosperm branches of the Tree of Life.

[opposite right]
Horsetail (*Equisetum*), is one of the earliest examples of plants that evolved after the conquest of land.

[below]
The cycad, *Cycas revoluta*, also known as sago palm.

PLANTS IN THE NON-FLOWERING SECTION

Equisetales
Equisetaceae (horsetail family)
Equisetum

Osmundales
Osmundaceae (royal fern family)
Osmunda

Cyatheales
Cyatheaceae (tree-fern family)
Dicksonia

Polypodiales
Pteridaceae (ribbon-fern family)
Adiantum

Aspleniaceae (spleenwort family)
Asplenium
Athyrium
Woodwardia

Polypodiaceae (polypody family)
Blechnum
Dryopteris
Lepisorus
Matteuccia
Polypodium
Polystichum

Cycadales
Cycadaceae (sago palm family)
Cycas

Ginkgoales
Ginkgoaceae (maidenhair tree family)
Ginkgo

Ephedrales
Ephedraceae (joint fir family)
Ephedra

Araucariales
Araucariaceae (kauri-tree family)
Wollemia

Podocarpaceae (yew pine family)
Podocarpus

Pinales
Pinaceae (pine family)
Pinus

Cupressales
Sciadopityaceae (umbrella-pine family)
Sciadopitys

Cupressaceae (cypress family)
Juniperus
Microbiota

Taxaceae (yew family)
Taxus

Magnoliids

It is unclear when the first flowering plants (angiosperms) emerged. Some scientists, using DNA evidence, have speculated it could have been as early as 245 million years ago but the oldest fossil flowers to have been found date from around 140 million years ago. A conservative estimate would be that the first flowers appeared around 155 million years ago, in the late Jurassic period..

Reproduction in early flowering plants required pollination of flowers by insects. The earliest flowers were simply constructed compared to more recently evolved flowers; consequently, pollination would occur whichever way insects entered the flower. These early flowers had either numerous large floral parts, often producing a strong, intoxicating scent, or a few small parts bearing nectar-producing glands. They all evolved to attract pollinators.

Many descendants of these early flowering plants are easily recognisable, and they often feature in our gardens and greenhouses today. Magnolias are one example. Flowers and their new relationships with pollinators heralded a period of co-evolution for insects and other animals and plants – a complex story that continues to unfold today.

Although outside the magnoliids clade (group), the order Austrobaileyales, which includes star anise (*Illicium*), is also represented in this garden room. This order diverged earlier than the magnoliids.

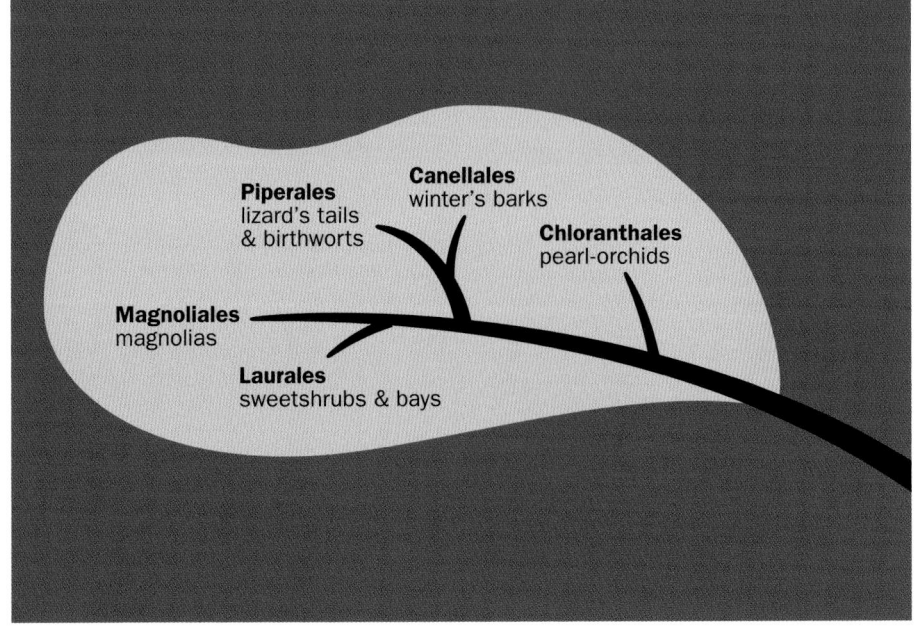

[left]
The magnoliids branch of the Tree of Life.

[opposite]
The first flowering plants evolved to attract early pollinators and their direct descendants include magnolias (top) and *Calycanthus* (below).

PLANTS IN THE MAGNOLIIDS SECTION

Austrobaileyales
Schisandraceae (star anise family)
  Illicium
  Kadsura
  Schisandra

Canellales
Winteraceae (winter's bark family)
  Drimys

Piperales
Saururaceae (lizard's tail family)
  Anemopsis
  Houttuynia

Aristolochiaceae (birthwort family)
  Asarum
  Aristolochia
  Saruma

Magnoliales
Magnoliaceae (magnolia family)
  Magnolia

Laurales
Calycanthaceae (spicebush family)
  Calycanthus

Chimonanthus
Lauraceae (bay laurel family)
  Laurus

Monocots

Monocots (monocotyledons) are flowering plants in which seeds usually contain only one embryonic leaf (cotyledon) at germination. This remarkable group provides 52% of the carbohydrates eaten by humans worldwide. Although there are approximately 60,000 species of monocots, we rely heavily on only three to feed us: maize, wheat and rice.

The domestication of these crops enabled us to become Earth's most dominant animal. By dedicating a third of all agricultural land to their cultivation, we have helped monocots to become Earth's most abundant plants in terms of land coverage. The monocots probably did not need our help to succeed because they are highly adaptable and extremely varied. Some palms, for example, are so large they are visible to satellites in space, and there are creeping vines that are two hundred metres long. Other monocots, such as tulips and daffodils, are geophytes living underground for much of the year whereas others, including many orchids, are epiphytes that cling to trees or rocks. As a result of such flexibility, monocots grow in every ecological zone on the planet, including marine and freshwater environments, deserts and even the edges of glaciers. The monocot lineage diverged from the Tree of Life around 150 million years ago during the late part of the Jurassic period.

[above]
Summer flowering *Lilium regale* can be found in the monocots garden room.

[left]
The monocots branch of the Tree of Life.

PLANTS IN THE MONOCOTS SECTION

Alismatales
Araceae (calla lily family)
  Arisaema
  Arisarum
  Arum
  Zantedeschia

Liliales
Melanthiaceae
(wake robin family)
  Trillium
  Veratrum

Alstroemeriaceae
(Inca lily family)
  Alstroemeria

Colchicaceae
(naked lady family)
  Colchicum

Liliaceae (lily family)
  Erythronium
  Fritillaria
  Lilium
  Tricyrtis
  Tulipa

Asparagales
Iridaceae (iris family)
  Crocosmia
  Crocus
  Diplarrhena

Gladiolus
Hesperantha
Iris
Libertia
Moraea
Sisyrinchium
Tritonia
Watsonia

Orchidaceae
(orchid family)
  Bletilla
  Epipactis

Asphodelaceae
(daylily family)
  Asphodeline
  Asphodelus
  Bulbine
  Dianella
  Eremurus
  Hemerocallis
  Kniphofia
  Phormium

Amaryllidaceae
(onion family)
  Agapanthus
  Allium
  Amaryllis
  Crinum
  Leucojum
  Narcissus

Nerine
Sternbergia
Tulbaghia

Asparagaceae
(hyacinth family)
  Camassia
  Cordyline
  Dasylirion
  Eucomis
  Hosta
  Hyacinthus
  Liriope
  Maianthemum
  Ophiopogon
  Ornithogalum
  Ruscus
  Scilla
  Yucca

Arecales
Arecaceae
(palm family)
  Jubaea
  Phoenix
  Trachycarpus

Commelinales
Commelinaceae
(spiderwort family)
  Commelina

Zingiberales
Zingiberaceae
(ginger family)
  Cautleya
  Hedychium
  Roscoea

Poales
Bromeliaceae
(pineapple family)
  Fascicularia

Juncaceae
(rush family)
  Juncus

Cyperaceae
(sedge family)
  Carex

Poaceae
(grass family)
  Calamagrostis
  Deschampsia
  Miscanthus
  Molinia
  Panicum
  Pennisetum
  Phaenosperma
  Phyllostachys
  Sporobolus

Early diverging eudicots

Eudicots began diverging from the main trunk of the Tree of Life around 140 million years ago during the Cretaceous period. This large group of plants (which includes the rosids and asterids) makes up 70% of the flowering plants. In this garden room are examples of plants that evolved early on in this radiation – called the early diverging eudicots.

The common characteristic that unites eudicots is the shape of their pollen. Described by botanists as 'tricolpate', this pollen features three grooves or pores. Its existence has been known for a long time, but the plant groups that produced it were so morphologically different that they hadn't previously been considered a closely related group. DNA analysis has uncovered the definite relationships between these plants at a molecular level and allowed the eudicots to be classified together for the first time. This is a good example of how advances in DNA technology have changed the way plants are grouped, reframing our understanding of plant evolution.

Within this garden you'll encounter representatives of early diverging or first evolving groups of eudicots including the orders Buxales (box), Proteales (protea and grevilleas) and Ranunculales (buttercups and poppies).

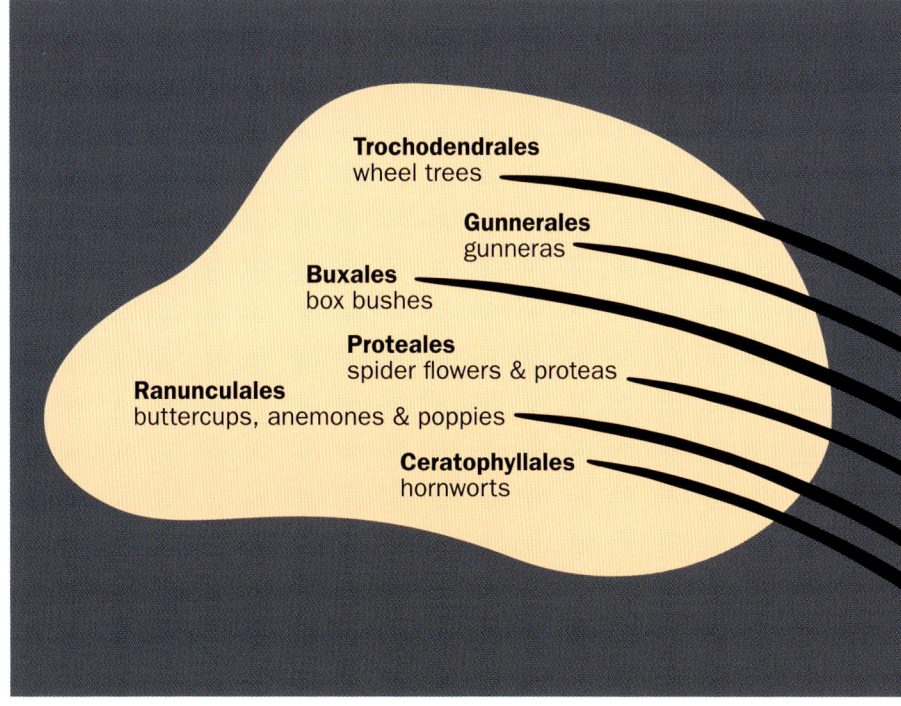

[left]
The branches of the Tree of Life representing the early diverging eudicots.

[opposite]
The Californian poppy, *Eschscholzia californica*, in the poppy family.

PLANTS IN THE EARLY DIVERGING EUDICOTS SECTION

Ranunculales
Papaveraceae
(poppy family)
   Argemone
   Chelidonium
   Corydalis
   Dendromecon
   Dicentra
   Eschscholzia
   Glaucium
   Macleaya
   Meconopsis
   Papaver
   Romneya
   Stylophorum

Ranunculaceae
(buttercup family)
   Aconitum
   Actaea
   Adonis
   Anemone
   Anemonopsis
   Aquilegia
   Caltha
   Clematis
   Delphinium
   Helleborus
   Nigella
   Pulsatilla
   Ranunculus
   Thalictrum
   Trollius

Lardizabalaceae
(zabala fruit family)
   Decaisnea

Berberidaceae
(barberry family)
   Berberis
   Epimedium
   Jeffersonia
   Mahonia
   Nandina
   Podophyllum

Proteales
Proteaceae
(sugarbush family)
   Grevillea
   Hakea
   Persoonia
   Telopea

Buxales
Buxaceae
(box family)
   Buxus
   Pachysandra
   Sarcococca

Gunnerales
Gunneraceae
(giant-rhubarb family)
   Gunnera

EARLY DIVERGING EUDICOTS   23

Rosids I

Rosids and asterids, two large sister clades, diverged from each other during the Cretaceous period, around 120 million years ago. Rosid families are very diverse, containing both herbaceous and woody plants. They dominate both our temperate and tropical forests, producing most of our high-canopy trees. Many rosids have flower parts in multiples of four or five like the hollyhocks (*Alcea),* rock roses (*Cistus*) and *Geranium*.

Rosids include some of our best-loved garden plants and also food plants, such as cabbages (*Brassica*) and sea kale (*Crambe*) in the mustard family (Brassicaceae). These contain mustard oils, giving them a peppery taste. With 70,000 species, over a third of the world's angiosperms (flowering plants) belong to the rosids. They are divided between two garden rooms in the Evolution Garden. In the first of these are some of the earliest plants to evolve within the clade, such as the porcelain berry (*Ampelopsis brevipedunculata*) in the grapevine family (Vitaceae) and families in the order Saxifragales, which includes peonies, saxifrages and sedums.

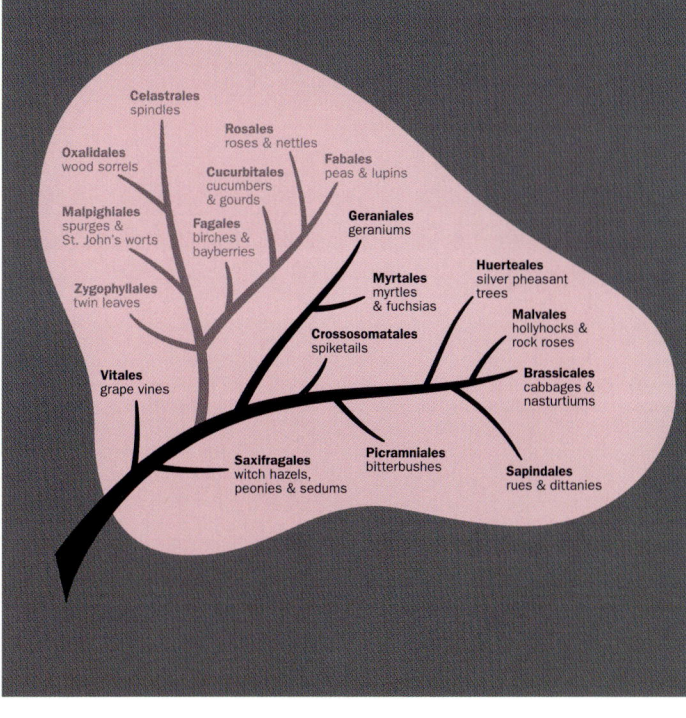

[above]
Geranium wallichianum is one of many good garden plants in the rosids clade.

[left]
The rosids branch of the Tree of Life. Plants in the first rosids garden room are selected from the orders highlighted.

PLANTS IN THE ROSIDS I SECTION

Saxifragales
Paeoniaceae
(peony family)
  Paeonia
  Peltaria

Hamamelidaceae
(witch hazel family)
  Hamamelis

Saxifragaceae
(saxifrage family)
  Astilbe
  Bergenia
  Chrysosplenium
  Heuchera
  Mukdenia
  Rodgersia
  Saxifraga
  Tellima
  Tiarella

Crassulaceae
(stonecrop family)
  Sedum

Vitales
Vitaceae (grapevine family)
  Ampelopsis

Geraniales
Geraniaceae
(crane's-bill family)
  Erodium
  Geranium

Francoaceae
(bridal-wreath family)
  Francoa
  Melianthus

Myrtales
Lythraceae
(pomegranate family)
  Lythrum
  Punicea

Onagraceae
(fuchsia family)
  Chamerion
  Clarkia
  Epilobium
  Fuchsia
  Gaura
  Oenothera

Myrtaceae
(myrtle family)
  Callistemon
  Leptospermum
  Melaleuca
  Ugni

Crossosomatales
Stachyuraceae
(spiketail family)
  Stachyurus

Sapindales
Anacardiaceae
(cashew family)
  Cotinus

Rutaceae
(citrus family)
  Choisya
  Correa
  Dictamnus
  Ruta
  Skimmia

Malvales
Malvaceae
(mallow family)
  Alcea
  Althaea
  Kitaibela
  Lavatera
  Malva
  Modiolastrum
  Sidalcea
  Sparrmannia

Thymelaeceae
(mezereon family)
  Daphne
  Edgeworthia

Cistaceae
(rock-rose family)
  Cistus
  Halimium
  Helianthemum

Brassicales
Tropaeolaceae
(nasturtium family
  Tropaeolum

Resedaceae
(mignonette family
  Reseda

Cleomaceae
(spider flower family)
  Cleome

Brassicaceae
(cabbage family)
  Aethionema
  Alyssum
  Aubrieta
  Aurinia
  Berteroa
  Brassica
  Cardamine
  Crambe
  Erysimum
  Fibigia
  Hesperis
  Iberis
  Isatis
  Lunaria
  Malcolmia
  Matthiola
  Pachyphragma
  Peltaria
  Sinapis

Rosids II

The second garden room for the rosids includes the rose family (Rosaceae). Many fruits produced by Rosaceae are edible, a characteristic that has been key to their success, as it means animals will eat them and help disperse the seeds. Cherry (*Prunus*) fruits, containing single hard stones, are a favourite for birds, along with many other berries in this family. Although roses (*Rosa*) are mostly cultivated for their beautiful flowers, the hips (fruits) have also been used for culinary and medicinal purposes for centuries.

Another important group of plants in the rosids is the legume family (Fabaceae). The legume family first evolved some 70–65 million years ago, during the Cretaceous period, a time of rapid insect diversification. For centuries, humans have cooked legume seeds to remove their harmful toxins and make them safe for us to eat. Lupin (*Lupinus*) seeds are a delicacy in many countries and are gaining recognition as a high-fibre, high protein, low-calorie crop. Lupin seeds also contain chemicals with anti-diabetic and anti-inflammatory properties.

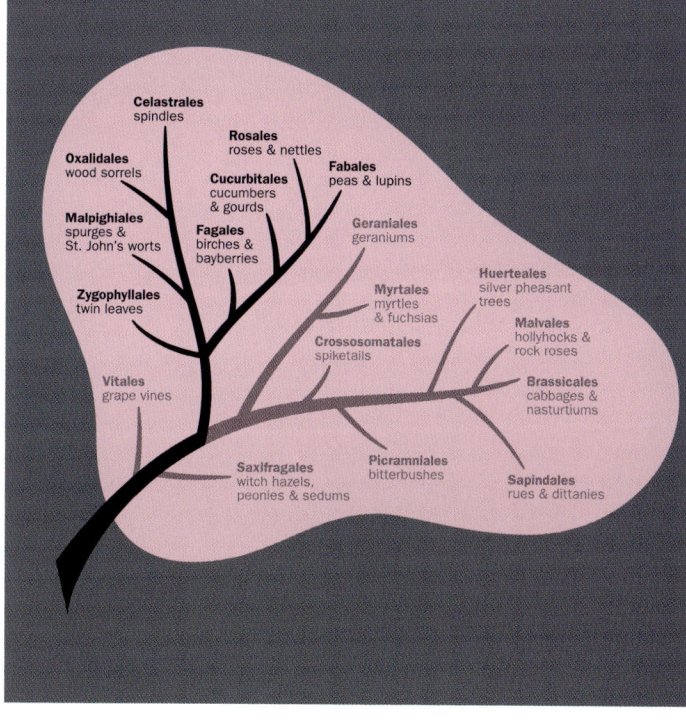

[above]
The hips of *Rosa glauca*.

[left]
The rosids branch of the Tree of Life. The plants in the second rosids garden room are selected from those highlighted.

EVOLUTION GARDEN

PLANTS IN THE ROSIDS II SECTION

Fabales
Fabaceae
(legume family)
  Amicia
  Anthyllis
  Apios
  Astragalus
  Baptisia
  Colutea
  Coronilla
  Cytisus
  Datisca
  Dorycnium
  Galega
  Genista
  Glycyrrhiza
  Hoita
  Lathyrus
  Lespedeza
  Lupinus
  Medicago
  Onobrychis
  Ononis
  Thermopsis
  Trifolium
  Vicia

Polygalaceae
(milkwort family)
  Polygala

Fagales
Myricaceae
(bayberry family)
  Myrica

Betulaceae
(birch family)
  Betula

Rosales
Rosaceae
(rose family)
  Acaena
  Agrimonia
  Alchemilla
  Aruncus
  Exochorda
  Filipendula
  Geum
  Gillenia
  Malus
  Potentilla
  Prunus
  Rosa
  Sanguisorba
  Sorbaria
  Spiraea

Rhamnaceae
(buckthorn family
  Ceanothus

Urticaceae
(nettle family)
  Boehmeria

Cucurbitales
Cucurbitaceae
(cucumber family)
  Bryonia
  Ecballium
  Sicyos
  Thladiantha

Datiscaceae
(durango root family)
  Datisca

Celastrales
Celastraceae
(spindle family)
  Euonymus

Oxalidales
Elaeocarpaceae
(fairy petticoat family)
  Crinodendron

Malpighiales
Hypericaceae
(St John's wort family)
  Hypericum

Violaceae
(violet family)
  Viola

Passifloraceae
(passionfruit family)
  Passiflora

Salicaceae
(willow family)
  Carrierea
  Salix

Euphorbiaceae
(spurge family)
  Euphorbia

Linaceae
(flax family)
  Linum

Asterids I

The asterids are a large clade of angiosperms within the broader grouping of the eudicots. In evolutionary history, asterids diverged from the rest of the eudicots relatively recently – approximately 120 million years ago, around the time the first primates appeared.

Asterids are mainly soft-stemmed (herbaceous) plants, and many evolved to fill niches in the woody rosid forests of the Cretaceous period. Generally, their flowers have only one plane of symmetry (zygomorphic), and many have fused sepals and petals. At the molecular level, various plants in this group contain bitter compounds, making them less tasty to herbivores and giving them a greater chance of survival. The asterids contain many of our important food crops, including potatoes and tomatoes. The toxicity of others, such as the deadly nightshade, makes them unsuitable for eating, although some produce potent chemical compounds that are hugely beneficial when used in medicines, like digitalin in the foxglove (*Digitalis*).

Among the plant families in the first garden room for the asterids are the tobacco family (Solanaceae), primrose family (Primulaceae), borage family (Boraginaceae) and mint family (Lamiaceae), which includes the sages (*Salvia*), the largest genus in Lamiaceae, with more than 900 species.

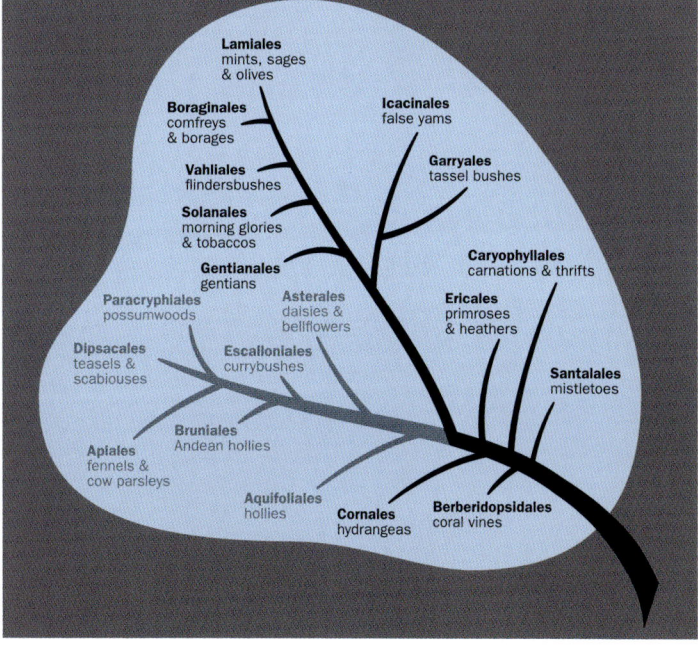

[above]
Primula bulleyana in the primrose family (Primulaceae).

[left]
The asterids branch of the Tree of Life. The plants in the first asterids garden room are selected from those highlighted.

PLANTS IN THE ASTERIDS I SECTION

Caryophyllales
Plumbaginaceae
(thrift family
  Armeria
  Ceratostigma
  Limonium
  Plumbago

Polygonaceae
(knotweed family)
  Bistorta
  Persicaria
  Rheum
  Rumex

Caryophyllaceae
(carnation family)
  Cerastium
  Dianthus
  Gypsophila
  Honckenya
  Lychnis
  Saponaria
  Silene
  Stellaria

Amaranthaceae
(spinach family)
  Suaeda

Phytolaccaceae
(pokeweed family)
  Phytolacca

Cornales
Hydrangeaceae
(hortensia family)
  Deutzia
  Hydrangea

Ericales
Polemoniaceae
(Jacob's ladder family)
  Phlox
  Polemonium

Primulaceae
(primrose family)
  Cyclamen
  Lysimachia
  Primula

Ericaceae (heather family)
  Erica
  Pieris

Gentianales
Rubiaceae (coffee family)
  Asperula
  Cephalanthus
  Galium
  Rubia

Gentianaceae
(gentian family)
  Gentiana

Apocynaceae
(periwinkle family
  Amsonia
  Vinca

Boraginales
Boraginaceae
(forget-me-not family)
  Anchusa
  Borago
  Echium
  Lindelofia
  Nonea
  Pulmonaria
  Symphytum
  Trachystemon

Solanales
Convolvulaceae
(bindweed family)
  Ipomoea

Solanaceae
(tobacco family)
  Fabiana
  Nicotiana
  Physalis

Lamiales
Oleaceae (olive family
  Forsythia
  Jasminum
  Olea

Plantaginaceae
(speedwell family)
  Chelone
  Digitalis
  Linaria
  Penstemon

  Plantago
  Veronica
  Veronicastrum

Scrophulariaceae
(figwort family)
  Verbascum

Acanthaceae
(bear's breeches family)
  Acanthus
  Brillantasia
  Strobilanthes

Verbenaceae
(vervain family)
  Verbena

Lamiaceae (mint family)
  Callicarpa
  Clinopodium
  Dorystaechas
  Lavandula
  Leonorus
  Marrubium
  Mentha
  Monarda
  Nepeta
  Origanum
  Perovskia
  Phlomis
  Physostegia
  Salvia
  Stachys
  Thymus

Asterids II

The second garden room for the asterids is dominated by one family – the daisy family (Asteraceae or Compositae), the largest family of flowering plants. It is a very widespread family, with all members sharing a common reproductive feature: the arrangement of individual florets into a 'false flower' (pseudanthium). This characteristic is easily recognisable in the plants displayed in this garden. What appears to be a single flower is actually composed of a number of smaller disc florets on a central disc, surrounded by a ring of ray florets resembling petals. Together this makes up a single flower head (capitulum). This adaptation has proven to be incredibly successful, as it increases the number of florets that can be pollinated in a single visit from a pollinator, and therefore the amount of seed produced.

Other families in this section include the bell flower family (Campanulaceae) and carrot family (Apiaceae). Many of the plants in Apiaceae have been used for medicinal and culinary purposes for centuries, such as cumin (*Cuminum cyminum*), anise (*Pimpinella anisum*), coriander (*Coriandrum sativum*), fennel (*Foeniculum vulgare*), dill (*Anethum graveolens*), carrots (*Daucus carota*) and parsnips (*Pastinaca sativa*).

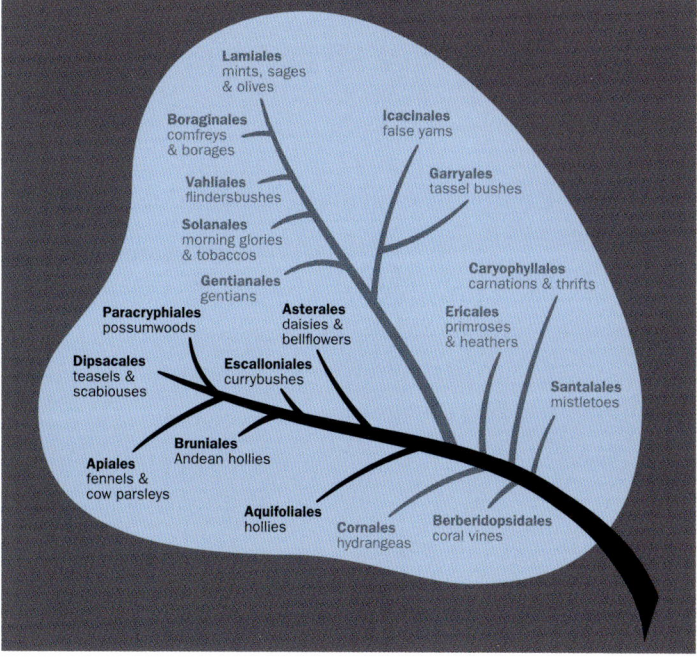

PLANTS IN THE ASTERIDS II SECTION

Aquifoliales
Helwingiaceae
(flowering-rafts family)
  Helwingia

Aquifoliaceae
(holly family)
  Ilex

Asterales
Campanulaceae
(bellflower family)
  Adenophora
  Campanula
  Lobelia
  Platycodon
  Pratia

Asteraceae
(daisy family)
  Achillea
  Anthemis
  Artemisia
  Aster
  Berkheya
  Centaurea
  Cirsium
  Cynara
  Dahlia
  Doronicum
  Echinacea
  Echinops
  Erigeron
  Eupatorium
  Eurybia

Helenium
Helianthus
Heliopsis
Inula
Liatris
Ligularia
Osteospermum
Ozothamnus
Ratibida
Rudbeckia
Santolina
Senecio
Silphium
Solidago
Stokesia
Symphyotrichum
Tanacetum
Vernonia

Dipsacales
Viburnaceae (elder family)
  Sambucus

Caprifoliaceae
(honeysuckle family)
  Centranthus
  Cephalaria
  Dipsacus
  Knautia
  Morina
  Patrinia
  Scabiosa
  Succisa
  Valeriana
  Weigela

Apiales
Pittosporaceae
(cheesewood family)
  Pittosporum

Araliaceae (ivy family)
  Schefflera

Apiaceae
(carrot family)
  Anthriscus
  Astrantia
  Bupleurum
  Cenolophium
  Chaerophyllum
  Crithmum
  Eryngium
  Foeniculum
  Molopospermum
  Peucadenum
  Pimpinella
  Selinum
  Zizia

[opposite left]
The asterids branch of the Tree of Life. The plants in the second asterids garden room are selected from those highlighted.

[opposite right]
Echinacea purpurea in the daisy family (Asteraceae or Compositae), the largest family of flowering plants.

ASTERIDS II

Acknowledgements

The Agius Evolution Garden opened on 3 July 2019.

The Board of Trustees of Royal Botanic Gardens, Kew would like to thank Marcus and Kate Agius for generously funding the creation of the Agius Evolution Garden.

The authors would like to thank the following for their input into this text: Harriet Stigner, Patrick Mulligan, Patrick Walsh, Nick Dent, Georgina Barley and Keith Johnson as well as Mark Chase, Mike Fay, Maarten Christenhusz, Vanessa Barber, Richard Barley and William Baker.

© The Board of Trustees of the Royal Botanic Gardens, Kew 2020

Illustrations and photographs Richard Wilford / © RBG Kew

The authors have asserted their right to be identified as the authors of this work in accordance with the Copyright, Designs and Patents Act 1988

All rights reserved. No part of this publication may be reproduced, stored in a retrieval system, or transmitted, in any form, or by any means, electronic, mechanical, photocopying, recording or otherwise, without written permission of the publisher unless in accordance with the provisions of the Copyright Designs and Patents Act 1988.

Great care has been taken to maintain the accuracy of the information contained in this work. However, neither the publisher nor the authors can be held responsible for any consequences arising from use of the information contained herein. The views expressed in this work are those of the authors and do not necessarily reflect those of the publisher or of the Board of Trustees of the Royal Botanic Gardens, Kew.

First published in 2020

Royal Botanic Gardens, Kew, Richmond, Surrey, TW9 3AB, UK

www.kew.org

Distributed on behalf of the Royal Botanic Gardens, Kew in North America by The University of Chicago Press, 1427 East 60th St, Chicago, IL 60637, USA.

ISBN 978 1 84246 710 7

British Library Cataloguing in Publication Data
A catalogue record for this book is available from the British Library

Copy-editing: Michelle Payne
Design and page layout: Ocky Murray
Production management: Georgina Hills, Jo Pillai

Printed in Italy by Printer Trento s. r. l.

For information or to purchase all Kew titles please visit shop.kew.org/kewbooksonline or email publishing@kew.org

Kew's mission is to be the global resource in plant and fungal knowledge, and the world's leading botanic garden.

Kew receives approximately one third of its funding from Government through the Department for Environment, Food and Rural Affairs (Defra). All other funding needed to support Kew's vital work comes from members, foundations, donors and commercial activities including book sales.